MY

FAVORITE RECIPES

Contents

Recipe Nº. 1:

Servings

Prep time

Cook time

Ingredients:

Directions:

Notes:

Recipe Nº. 2:

Servings

Prep time

Cook time

Ingredients:

Directions:

Notes:

Recipe Nº. 3:

Servings

Prep time

Cook time

Ingredients:

Directions:

Notes:

Recipe Nº. 4:

Servings

Prep time

Cook time

Ingredients:

Directions:

Notes:

Recipe N°. 5:

Servings

Prep time

Cook time

Ingredients:

Directions:

Notes:

Recipe Nº. 6:

Servings

Prep time

Cook time

Ingredients:

Directions:

Notes:

Recipe Nº. 7:

Servings

Prep time

Cook time

Ingredients:

Directions:

Notes:

Recipe N°. 8:

Servings

Prep time

Cook time

Ingredients:

Directions:

Notes:

Recipe Nº. 9:

Servings

Prep time

Cook time

Ingredients:

Directions:

Notes:

Recipe Nº. 10:

Servings

Prep time

Cook time

Ingredients:

Directions:

Notes:

Recipe N°. 11:

Servings

Prep time

Cook time

Ingredients:

Directions:

Notes:

Recipe Nº. 12:

Servings

Prep time

Cook time

Ingredients:

Directions:

Notes:

Recipe Nº. 13:

Servings

Prep time

Cook time

Ingredients:

Directions:

Notes:

Recipe Nº. 14:

Servings

Prep time

Cook time

Ingredients:

Directions:

Notes:

Recipe Nº. 15:

Servings

Prep time

Cook time

Ingredients:

Directions:

Notes:

Recipe Nº. 16:

Servings

Prep time

Cook time

Ingredients:

Directions:

Notes:

Recipe Nº. 17:

Servings

Prep time

Cook time

Ingredients:

Directions:

Notes:

Recipe Nº. 18:

Servings

Prep time

Cook time

Ingredients:

Directions:

Notes:

Recipe Nº. 19:

Servings

Prep time

Cook time

Ingredients:

Directions:

Notes:

Recipe Nº. 20:

Servings

Prep time

Cook time

Ingredients:

Directions:

Notes:

Recipe Nº. 21:

Servings

Prep time

Cook time

Ingredients:

Directions:

Notes:

Recipe Nº. 22:

Servings

Prep time

Cook time

Ingredients:

Directions:

Notes:

Recipe Nº. 23:

Servings

Prep time

Cook time

Ingredients:

Directions:

Notes:

Recipe Nº. 24:

Servings

Prep time

Cook time

Ingredients:

Directions:

Notes:

Recipe Nº. 25:

Servings

Prep time

Cook time

Ingredients:

Directions:

Notes:

Recipe Nº. 26:

Servings

Prep time

Cook time

Ingredients:

Directions:

Notes:

Recipe Nº. 27:

Servings

Prep time

Cook time

Ingredients:

Directions:

Notes:

Recipe Nº. 28:

Servings

Prep time

Cook time

Ingredients:

Directions:

Notes:

Recipe Nº. 29:

Servings

Prep time

Cook time

Ingredients:

Directions:

Notes:

Recipe Nº. 30:

Servings

Prep time

Cook time

Ingredients:

Directions:

Notes:

Recipe Nº. 31:

Servings

Prep time

Cook time

Ingredients:

Directions:

Notes:

Recipe Nº. 32:

Servings

Prep time

Cook time

Ingredients:

Directions:

Notes:

Recipe N°. 33:

Servings

Prep time

Cook time

Ingredients:

Directions:

Notes:

Recipe Nº. 34:

Servings

Prep time

Cook time

Ingredients:

Directions:

Notes:

Recipe Nº. 35:

Servings

Prep time

Cook time

Ingredients:

Directions:

Notes:

Recipe N°. 36:

Servings

Prep time

Cook time

Ingredients:

Directions:

Notes:

Recipe Nº. 37:

Servings

Prep time

Cook time

Ingredients:

Directions:

Notes:

Recipe Nº. 38:

Servings

Prep time

Cook time

Ingredients:

Directions:

Notes:

Recipe Nº. 39:

Servings

Prep time

Cook time

Ingredients:

Directions:

Notes:

Recipe Nº. 40:

Servings

Prep time

Cook time

Ingredients:

Directions:

Notes:

Recipe Nº. 41:

Servings

Prep time

Cook time

Ingredients:

Directions:

Notes:

Recipe Nº. 42:

Servings

Prep time

Cook time

Ingredients:

Directions:

Notes:

Recipe Nº. 43:

Servings

Prep time

Cook time

Ingredients:

Directions:

Notes:

Recipe Nº. 44:

Servings

Prep time

Cook time

Ingredients:

Directions:

Notes:

Recipe Nº. 45:

Servings

Prep time

Cook time

Ingredients:

Directions:

Notes:

Recipe Nº. 46:

Servings

Prep time

Cook time

Ingredients:

Directions:

Notes:

Recipe Nº. 47:

Servings

Prep time

Cook time

Ingredients:

Directions:

Notes:

Recipe Nº. 48:

Servings

Prep time

Cook time

Ingredients:

Directions:

Notes:

Recipe Nº. 49:

Servings

Prep time

Cook time

Ingredients:

Directions:

Notes:

Recipe Nº. 50:

Servings

Prep time

Cook time

Ingredients:

Directions:

Notes:

Recipe Nº. 51:

Servings

Prep time

Cook time

Ingredients:

Directions:

Notes:

Recipe Nº. 52:

Servings

Prep time

Cook time

Ingredients:

Directions:

Notes:

Recipe Nº. 53:

Servings

Prep time

Cook time

Ingredients:

Directions:

Notes:

Recipe Nº. 54:

Servings

Prep time

Cook time

Ingredients:

Directions:

Notes:

Recipe Nº. 55:

Servings

Prep time

Cook time

Ingredients:

Directions:

Notes:

Recipe No. 56:

Servings

Prep time

Cook time

Ingredients:

Directions:

Notes:

Recipe Nº. 57:

Servings

Prep time

Cook time

Ingredients:

Directions:

Notes:

Recipe Nº. 58:

Servings

Prep time

Cook time

Ingredients:

Directions:

Notes:

Recipe Nº. 59:

Servings

Prep time

Cook time

Ingredients:

Directions:

Notes:

Recipe Nº. 60:

Servings

Prep time

Cook time

Ingredients:

Directions:

Notes:

Recipe N°. 61:

Servings

Prep time

Cook time

Ingredients:

Directions:

Notes:

Recipe Nº. 62:

Servings

Prep time

Cook time

Ingredients:

Directions:

Notes:

Recipe Nº. 63:

Servings

Prep time

Cook time

Ingredients:

Directions:

Notes:

Recipe Nº. 64:

Servings

Prep time

Cook time

Ingredients:

Directions:

Notes:

Recipe Nº. 65:

Servings

Prep time

Cook time

Ingredients:

Directions:

Notes:

Recipe Nº. 66:

Servings

Prep time

Cook time

Ingredients:

Directions:

Notes:

Recipe Nº. 67:

Servings

Prep time

Cook time

Ingredients:

Directions:

Notes:

Recipe Nº. 68:

Servings

Prep time

Cook time

Ingredients:

Directions:

Notes:

Recipe Nº. 69:

Servings

Prep time

Cook time

Ingredients:

Directions:

Notes:

Recipe Nº. 70:

Servings

Prep time

Cook time

Ingredients:

Directions:

Notes:

Recipe Nº. 71:

Servings

Prep time

Cook time

Ingredients:

Directions:

Notes:

Recipe Nº. 72:

Servings

Prep time

Cook time

Ingredients:

Directions:

Notes:

Recipe Nº. 73:

Servings

Prep time

Cook time

Ingredients:

Directions:

Notes:

Recipe Nº. 74:

Servings

Prep time

Cook time

Ingredients:

Directions:

Notes:

Recipe Nº. 75:

Servings

Prep time

Cook time

Ingredients:

Directions:

Notes:

Recipe Nº. 76:

Servings

Prep time

Cook time

Ingredients:

Directions:

Notes:

Recipe No. 77:

Servings

Prep time

Cook time

Ingredients:

Directions:

Notes:

Recipe Nº. 78:

Servings

Prep time

Cook time

Ingredients:

Directions:

Notes:

Recipe N°. 79:

Servings

Prep time

Cook time

Ingredients:

Directions:

Notes:

Recipe Nº. 80:

Servings

Prep time

Cook time

Ingredients:

Directions:

Notes:

Recipe Nº. 81:

Servings

Prep time

Cook time

Ingredients:

Directions:

Notes:

Recipe Nº. 82:

Servings

Prep time

Cook time

Ingredients:

Directions:

Notes:

Recipe Nº. 83:

Servings

Prep time

Cook time

Ingredients:

Directions:

Notes:

Recipe No. 84:

Servings

Prep time

Cook time

Ingredients:

Directions:

Notes:

Recipe N°. 85:

Servings

Prep time

Cook time

Ingredients:

Directions:

Notes:

Recipe Nº. 86:

Servings

Prep time

Cook time

Ingredients:

Directions:

Notes:

Recipe Nº. 87:

Servings

Prep time

Cook time

Ingredients:

Directions:

Notes:

Recipe Nº. 88:

Servings

Prep time

Cook time

Ingredients:

Directions:

Notes:

Recipe Nº. 89:

Servings

Prep time

Cook time

Ingredients:

Directions:

Notes:

Recipe Nº. 90:

Servings

Prep time

Cook time

Ingredients:

Directions:

Notes:

Recipe N°. 91:

Servings

Prep time

Cook time

Ingredients:

Directions:

Notes:

Recipe Nº. 92:

Servings

Prep time

Cook time

Ingredients:

Directions:

Notes:

Recipe N°. 93:

Servings

Prep time

Cook time

Ingredients:

Directions:

Notes:

Recipe Nº. 94:

Servings

Prep time

Cook time

Ingredients:

Directions:

Notes:

Recipe Nº. 95:

Servings

Prep time

Cook time

Ingredients:

Directions:

Notes:

Recipe Nº. 96:

Servings

Prep time

Cook time

Ingredients:

Directions:

Notes:

Recipe Nº. 97:

Servings

Prep time

Cook time

Ingredients:

Directions:

Notes:

Recipe Nº. 98:

Servings

Prep time

Cook time

Ingredients:

Directions:

Notes:

Recipe Nº. 99:

Servings

Prep time

Cook time

Ingredients:

Directions:

Notes:

Recipe Nº. 100:

Servings

Prep time

Cook time

Ingredients:

Directions:

Notes:

Recipe Nº. 101:

Servings

Prep time

Cook time

Ingredients:

Directions:

Notes:

Recipe Nº. 102:

Servings

Prep time

Cook time

Ingredients:

Directions:

Notes:

Recipe Nº. 103:

Servings

Prep time

Cook time

Ingredients:

Directions:

Notes:

Recipe No. 104:

Servings

Prep time

Cook time

Ingredients:

Directions:

Notes:

Recipe Nº. 105:

Servings

Prep time

Cook time

Ingredients:

Directions:

Notes:

Recipe Nº. 106:

Servings

Prep time

Cook time

Ingredients:

Directions:

Notes:

Recipe Nº. 107:

Servings

Prep time

Cook time

Ingredients:

Directions:

Notes:

Recipe Nº. 108:

Servings

Prep time

Cook time

Ingredients:

Directions:

Notes:

Recipe Nº. 109:

Servings

Prep time

Cook time

Ingredients:

Directions:

Notes:

Recipe Nº. 110:

Servings

Prep time

Cook time

Ingredients:

Directions:

Notes:

Recipe Nº. 111:

Servings

Prep time

Cook time

Ingredients:

Directions:

Notes:

Recipe N°. 112:

Servings

Prep time

Cook time

Ingredients:

Directions:

Notes:

Recipe Nº. 113:

Servings

Prep time

Cook time

Ingredients:

Directions:

Notes:

Recipe Nº. 114:

Servings

Prep time

Cook time

Ingredients:

Directions:

Notes:

Recipe Nº. 115:

Servings

Prep time

Cook time

Ingredients:

Directions:

Notes:

Recipe Nº. 116:

Servings

Prep time

Cook time

Ingredients:

Directions:

Notes:

Recipe Nº. 117:

Servings

Prep time

Cook time

Ingredients:

Directions:

Notes:

Recipe Nº. 118:

Servings

Prep time

Cook time

Ingredients:

Directions:

Notes:

Recipe Nº. 119: ....

.... Servings

.... Prep time

.... Cook time

Ingredients:

Directions:

Notes:

Recipe Nº. 120:

Servings

Prep time

Cook time

Ingredients:

Directions:

Notes:

Notes

Printed by Amazon Media EU S.à r.l., 5 Rue Plaetis, L-2338, Luxembourg

Images by: Oxy_gen (Cover) / Shutterstock.com

Fonts designed by: ratticsassin makes Fonts, / daFont.com (book title); Ageng Jatmiko / letterhend.com (text)

Made in the USA
Columbia, SC
26 April 2020